BOX TURTLE AS PETS

Complete owners guide to box turtle
training, care, reproduction, management
and many more included

ALBERT A. NELSON

Table of Contents

Introduction

Greetings from the wonderful box turtle friendship! This comprehensive guide will help you understand the subtleties of caring for these amazing reptiles that make them unique and valuable pets. Knowing what your box turtle needs for maintenance can help you provide a happy and healthy home, whether you're a first-time pet owner or a seasoned reptile enthusiast.

In this guide, we'll cover the many aspects of caring for box turtles, starting with the critical decision to create the perfect home. Learn how to build a comfortable and realistic-looking home that provides everything your box turtle needs to live.

Analyze the nutritional needs of box turtles and prepare the optimal nutrient ratio to maintain their health and vigor. We also cover common health problems and preventative measures so you can manage your Shell covered companion care yourself.

Building a relationship with your box turtle is rewarding, and we offer advice on how to deal with them in a way that builds trust and strengthens your bond. To ensure your box turtle lives a happy and healthy life, look for enrichment activities that provide both physical and cerebral stimulation.

Explore the social aspects of box turtle behavior to learn about their unique personalities and how they interpret their communication signals. We take an in-depth look at the different box turtle species, providing useful information to help you choose the right companion.

We'll continue with grooming tips along the way to keep your box turtle's shell colorful and healthy. Learn about the different stages of their life cycle from infancy to adulthood to gain a deeper appreciation for these amazing creatures.

Finally, we answer frequently asked questions, debunk myths, and provide helpful solutions to common problems. Along with your box turtle, get ready for an exciting journey of discovery and friendship as we go over the principles of ethical ownership in this comprehensive guide.

Chapter 1

Choosing the right setting: Building a comfortable home for your box turtle

Because of their terrestrial lifestyle, box turtles require a carefully designed environment to mimic their native habitat. In this in-depth analysis, we look at the important components of creating an ideal habitat, considering variables such as enclosure size, substrate, temperature, lighting, and enrichment. Making sure your box turtle has a suitable habitat is important to its overall happiness and well-being.

Understanding Natural Habitat: Before delving into the details of artificial habitat, it is important to understand the box turtle's natural habitat. These reptiles live in a variety of habitats such as wetlands, grasslands, and forests. They typically look for sheltered burrows as well as areas that provide shade and sunlight. By keeping these things in mind, we can create a suitable habitat for the safety of our box turtles.

Enclosure Size and Design: The enclosure plays an important role in keeping your box turtle healthy. A large enclosure allows for increased movement and more natural features. For an adult box turtle, you should provide a habitat that is at least four feet by four feet. If you plan to keep more than one turtle, you should adjust the size of the enclosure.

Choose materials such as wood or hard plastic for the walls to create a safe and durable home. Adequate ventilation is necessary to maintain healthy air circulation, which will not increase humidity and possibly cause breathing problems.

A combination of cypress mulch, coconut husks and organic topsoil works well for the desired texture and moisture retention. Digging and burrowing are examples of natural behaviors that can be encouraged by selecting appropriate substitutes to replicate the natural environment. Your box turtle must be able to burrow to perform thermoregulation and nesting behaviors, so the substrate must be deep enough to do so.

Thermal Efficiency: The safety of your box turtle depends on the enclosure's ability to maintain an appropriate temperature gradient. Since these reptiles are ectothermic, external heat sources regulate their body temperature. Maintain a temperature of 90 to 95 degrees F (32 to 35 degrees C) for the oven and 21 to 24 degrees C (70 to 75 degrees F) for the refrigerator. Tank heating pads or heating bulbs should be used to achieve and maintain these heat zones.

UVB Lighting: Because box turtles absorb UVB rays in their natural habitat, it is important to replicate this in captivity with UVB-emitting lights. To ensure maximum efficiency, make sure the bulbs provide the appropriate UVB spectrum and replace them as recommended by the manufacturer. Box turtles need exposure to UVB light to properly assimilate calcium and maintain the health of their shells and bones.

Hiding and Enriching Areas: Box turtles like to explore their environment, so in addition to hiding in half logs, bushes or other places that serve as shelter, provide objects to climb and explore. This will help your box turtle feel less stressed and more secure.

Enrichment is important for your box turtle's mental and physical well-being. You should introduce objects such as branches, logs and rocks that stimulate their natural curiosity and behavior. Consider rotating or moving these pieces often to create a dynamic and engaging environment.

Water Source: A clean, readily available source of water is essential to your box turtle's overall health and hydration. Provide enough shallow water for drinking and diving. Change the water regularly to reduce contamination. Control the humidity to prevent the enclosure from becoming too wet.

Regular habitat care ensures that your box turtle lives in a clean and healthy environment. To prevent the growth of bacteria and parasites, dispose of garbage, uneaten food and any waste regularly. Thoroughly clean the enclosure once a month, replacing the mat and cleaning any areas that need it.

Monitoring Health and Behavior: Monitoring your box turtle's behavior is important to detect any health problems early. You should watch for changes in appearance or signs of fatigue, as well as eating and baking habits. Schedule regular check-ups with your veterinarian and seek medical attention as soon as you notice any problems.

In conclusion, providing your box turtle with a comfortable and suitable home is a fun project that immediately improves the quality of life. Your box turtle will thrive in captivity if you can understand and mimic its natural habitat in addition to providing suitable temperatures, UVB lighting, hiding places, and enrichment. Appropriate ownership includes regular monitoring, maintenance and veterinary care to ensure a happy and lasting relationship with these amazing reptiles. Remember that the foundation of a happy and healthy box turtle life is a well-designed habitat.

Chapter 2

Feeding the Turtle: Dietary Requirements and Dietary Guidelines for Box Turtles

The long-term health of your box turtle depends on feeding a healthy, balanced diet. This overview of feeding habits covers what box turtles naturally eat, what nutrients they need, and practical feeding tips to help keep your furry friend healthy.

Understanding Natural Diet: Box turtles are omnivorous and naturally consume a wide variety of plant and animal materials. This can include various fruits, vegetables, worms, slugs, snails and insects. Replenishment of essential nutrients in captivity is essential for box turtles to maintain good health.

Nutrients:

1. Protein: An important part of a box turtle's diet, protein helps the turtle's muscles and shell expand. Earthworms, mealworms, slugs and crickets are good sources of protein. Provide a variety of protein sources to ensure a balanced diet.

2. Vegetables: Add carrots, bell peppers, and squash in addition to dark, leafy greens like kale, collards, and dandelion greens to provide variety. To make the vegetables easier to eat, peel or cut them into bite-size pieces.

3. Fruits: Because of their high sugar content, fruits should be served only occasionally. As treats, you can include berries, melons and a small amount of apples or pears. Avoid unripe fruit to prevent spoilage.

4. Calcium: To increase the calcium content, chopped bone or some calcium powder is sprinkled on the food.

Exposure to UVB light helps the body absorb calcium. Shell growth and bone health are generally dependent on calcium.

5. Commercial diets: Using high-quality commercial turtle pellets or specially formulated foods is an easy way to ensure that box turtles have a balanced nutritional profile. Fresh foods should be added to the diet, they are not the only source of energy.

6. Feeding Frequency: The frequency of feeding your box turtle depends on its age and activity level. Adult box turtles can be fed every two or three days, but juveniles may need to be fed every day to grow. By monitoring their behavior and making adjustments to the feeding plan as needed, you can make sure your turtles are getting enough nutrients without going overboard.

7. Moisture: Box turtles need shallow clean water to stay well hydrated; It is useful to have access to a large container because you may want to put some of them inside. It is important to change the water frequently to prevent infection.

Nutrition guide:

1. Diversity is critical.
Offering a variety of foods not only allows box turtles to receive a variety of nutrients, but also helps mimic the species' natural diet. Experiment with different fruits, vegetables, and protein sources to keep their meals varied and nutritious.

2. Gut-loading insects: If you feed insects, consider feeding more nutritious foods to your box turtle before feeding them. This increases the nutritional value of the insects, which increases the turtle's diet.

3. Check Calcium and Phosphorus Ratio: It is very important to maintain proper calcium and phosphorus ratio. Eating foods high in phosphorus (such as certain insects or meats) can cause calcium deficiency and metabolic bone disease.

4. Seasonal Fluctuations: Box turtles can show seasonal fluctuations in hunger. During the colder months, your metabolism slows down, so you may eat less. To ensure their continued health, adjust how often they eat and monitor their weight.

5. Watch your chewing habits: Cut or chop vegetables into small pieces to make eating and digestion easier. Box turtles' beaks are designed for tearing, not grinding.

6. Avoid toxic foods: Foods that can be toxic to box turtles include onions, garlic, and high oxalate greens

like spinach. Learn which foods are toxic and which foods are acceptable to prevent unintended harm.

Nutrition Tips for Different Life Stages:

1. Hatchlings: A serving of finely chopped vegetables and small tender spines will develop their small stature and jaws. Chicks have different nutritional needs than adults, focusing on protein for growth.

2. Adults: As they get older, box turtles should have a varied diet that includes vegetables and protein. Adjust the feeding schedule to take into account their frozen metabolic rate.

to sum up:
Carefully balance protein, fruit, vegetables and supplements to ensure your box turtle gets the nutrition it needs. You can learn about your box turtle's natural

diet, offer a variety of foods, and consider age-specific needs to ensure your turtle has a balanced and healthy diet. Responsible turtle ownership involves regular behavior and weight monitoring, as well as making necessary adjustments to feeding schedules. By being attentive to details and committed to their diet, you can extend the life and health of your box turtle companion.

Chapter 3

Box turtle health: common issues and preventative measures

Being a good owner means taking care of your box turtle. We'll discuss common health issues that box turtles face, preventative measures you can take to keep them healthy, and the importance of regular veterinary care.

Identify common health issues:

1. Respiratory infections

Lethargy, wheezing, and runny nose can be signs of respiratory infections in box turtles, which are often caused by low humidity or temperature. This can be prevented by ensuring that the enclosure has adequate ventilation and a well-controlled temperature.

2. Rotten shells:

Shell rot A fungal or bacterial infection that damages the shell; It usually occurs due to prolonged exposure to moisture or dirty environment and soft spots, discoloration or unpleasant odor appear on the shell. Careful prevention and treatment of shell rot is important, regular shell inspections and maintenance of a clean, dry habitat are essential.

3. Metabolic bone disease or MBD:

An imbalance in the calcium-phosphorus ratio in the diet or insufficient UVB exposure can lead to metabolic bone disease, which weakens bones and causes shell abnormalities. Important preventative measures include calcium supplementation, UVB exposure, and ensuring a balanced diet.

4. Parasitic infection;

In addition to eating a healthy diet and maintaining good hygiene, prevention of parasite problems includes regular fecal examinations by the veterinarian. Box turtles may show signs of internal or external parasites such as loss of appetite, lethargy, and diarrhea.

5. Ovulation by women:

Egg binding is a problem female box turtles face when laying eggs. The cause, for example, is that there is not enough calcium in their diet or their nests. Providing a suitable nest site with suitable substrate and ensuring that the diet is rich in calcium will help prevent egg attachment.

6. Lack of fluid;

Box turtles are often concerned about dehydration, which can be insufficient water or a habitat that is too dry. Providing a shallow bowl of water and monitoring your turtle's humidity are important preventive measures. You can also submerge your turtle in water occasionally to keep it hydrated.

Interventions in preventive care;

1. Properly designed housing:

Establishing and maintaining suitable habitat is a cornerstone of preventive care. Make sure the enclosure has a clean surface, proper humidity and temperature gradient. Regular cleaning prevents bacteria and parasites from forming.

2. Sufficient light

Let your box turtle enjoy natural sunlight whenever possible, but be aware that it can overheat. The overall health of the box turtle depends on exposure to UVB radiation because it promotes healthy calcium metabolism. Buy premium UVB bulbs and make sure to change according to the manufacturer's instructions.

3. Balanced diet

Eating a balanced diet is important to prevent malnutrition. Make sure your box turtle receives a variety of fruits, vegetables, and meats to ensure he's getting all the nutrients he needs. You can also supplement your diet with calcium to help maintain healthy bones and bones.

4. Routine medical examinations;

A qualified veterinarian can perform stool tests, health assessments, and potential concerns. Early detection is the key to effective treatment of health conditions. Schedule regular checkups with a veterinarian who specializes in reptile care.

5. Control of water supply;

Proper watering of your box turtle is important to prevent dehydration. Make sure your turtle always has access to fresh, clean water. If your turtle shows signs of dehydration, including dry eyes or feeling lethargic, you may want to keep them in shallow water to keep them hydrated.

6. Use of monitoring

Carefully check your box turtle's eating habits, general appearance, and behavior. Any behavioral changes, weight loss, or abnormalities in the shell may indicate underlying medical issues. Consult a veterinarian immediately if you have any concerns.

7. Prohibited New Items:-

If you are adding a new box turtle to your collection, you may want to consider isolating it. This helps prevent diseases or parasites from spreading to existing turtle populations. During quarantine, be sure to monitor your new turtle closely for any signs of illness.

8. Safe Work Practices:-

Take good care of your box turtle to reduce stress and potential injuries. Stress weakens the immune system, making the turtle more susceptible to disease. Hold the turtle gently and support its whole body when necessary.

Emergency response

Even with the best precautions, emergencies can happen, so it's important to know what to do in the event of a medical emergency.

1. Difficulty breathing;
If the turtle shows signs of respiratory problems such as difficulty breathing, consult a veterinarian immediately; Keep it aside and keep it in a warm place. Breathing problems can progress quickly.

2. Shell Damage:-

To determine the best course of action and assess the extent of the damage, seek emergency veterinary care if the turtle has a shell injury. Do not try to treat the turtle yourself.

3. Lack of appetite and apathy;

If your box turtle shows lethargy or lack of food, it may indicate an underlying health problem. Put the turtle down, monitor its behavior and consult a veterinarian.

to sum up:

Remember that the key to ensuring your box turtle companion lives a long and healthy life in captivity is to provide active care. Keeping your box turtle healthy requires early veterinary treatment, frequent monitoring and preventive measures. By recognizing common health issues, designing their habitat properly, feeding them a balanced diet, and responding promptly to concerns, you can help your box turtle companion live a long and healthy life.

Chapter 4

Handling Precautions: Develop a relationship and trust with your box turtle

One of the most rewarding aspects of having a reptile as a pet is getting to know your box turtle. Box turtles are by nature solitary and a bit shy, unlike larger breeds, building trust with them requires patience, tolerance, understanding and respect for their natural tendencies. In this in-depth analysis, we'll explore caring for box turtles, the importance of building trust, and tips to help you and your furry friend have a happy and fulfilling relationship.

Understanding the nature of box turtles:

Box turtles are not large creatures like dogs or cats and generally prefer isolation, so handling them regularly can stress them out as they are shy by nature. It is important to know that not all box turtles accept handling and cannot respect their privacy needs.

1. Patience is key:

With a box turtle, confidence is slowly gained. Patience is an important virtue in relationship building. Do not try to catch your turtle immediately; Instead, give him time to adjust to the new environment. Pay attention to how they work, look at their comfort zones and respect their privacy needs.

2. Gradual Introduction:

Gradual exposure increases comfort level and reduces the likelihood of negative attachment to handling. Start with short sessions and let the turtle explore the environment and your hand at its own pace. Avoid sudden movements or hasty attempts to pick up the turtle immediately, as this can be stressful.

3. Getting to know someone by hand:

Before you try to handle your box turtle, give them some time to get to know you. Spend some time talking gently and carefully approach their enclosure so that the turtle associates your presence with a safe haven.

4. The right time:

Choose the right time to handle: Box turtles are most active during the warm part of the day. When you sleep naturally, avoid waking up in the middle of the night or early in the morning. When management is proactive and active in nature, positive interactions are more likely to occur.

5. Recognize the signs of stress:

Stress symptoms are important to understand. Box turtles may show signs of stress by retreating into their shells, breathing rapidly, hissing, or exhibiting a significant withdrawal response. If you see any of these signs, immediately return the turtle to its enclosure and allow it to hide.

6. Support Management System:-

When you think your box turtle is ready to be handled, approach it slowly and carefully from your side, making sure to keep a firm grip to prevent accidental slips, and use both hands to lift the turtle so that its legs are not too compressed and strapped. The torso is supported. Box turtles are not very agile and can be hurt by even minor mishaps.

7. Quick but inspiring meetings:

After a few short handling sessions, as your turtle gets used to being held, gradually increase the duration of the sessions. Concluding the sessions by politely returning the turtle to the enclosure helps build trust.

8. Body language is important:

Box turtles communicate with their body language, so watch their reactions when you touch them. If they come out of their shell or show distress, it's important to respect their boundaries and return them to their nest.

9. Create a configuration for safe handling:

A calm and controlled environment promotes a positive experience for you and your box turtle. Avoid loud noises, sudden movements, and other potential distractions when handling your turtle.

10. Sweets and Behavioral Incentives:

Moderation is key to preventing overeating, as treats can foster a positive relationship between treats and treats. Small, turtle-friendly treats can be offered during and after handling to establish pleasant interactions.

11. Consistency in communication;

If you want to gradually build trust, you have to be consistent in your relationship. Box turtles appreciate consistent nesting routines, so regular gentle handling sessions are important, as well as monitoring the turtle's comfort level.

12. Inclusion of improvement

Enrichment improves positive relationships with handling and mental stimulation; To keep your box turtle's mind engaged during routine handling, consider placing it in a safe, controlled outdoor enclosure or exploring safe toys.

13. Respect each person's unique personality:

Each box turtle has individual characteristics. Some may need a little contact all the time, while others may be more tolerant of handling. Respect their decisions and recognize that intimacy can take many forms.

14. See common features:

Allowing your box turtle to behave normally when you meet it—whether it's crawling, sleeping in your arms, or just exploring—will help you both understand and adapt to its normal behaviors.

15. Check out the signs of happiness:

Positive signs indicate that handling your box turtle is starting to feel more natural. Some examples of these gestures can be as basic as taking a comfortable position, exploring at a leisurely pace, or stretching the legs.

16. Outdoor inspection;

Building a safe outdoor habitat with plenty of sunlight allows your box turtle to enjoy outdoor exploration. Time outdoors provides stimulation and a change of scenery.

17. Routine examinations on health;

Use handling sessions to conduct brief health checks. Look for signs of discharge in the mouth, nose and eyes. Check the shell for damage or damage. Regular health check-ups can detect potential problems early.

18. Use the following when getting veterinary treatment

It may be necessary to treat patients frequently while receiving veterinary treatment. Your box turtle should adapt to handling gradually, especially if it requires medical attention. To help them prepare for the vet visit, it's important to handle them calmly and consistently in non-stressful situations.

to sum up:

Trusting and forming a deep bond with your box turtle will take time, but it will be worth it. To build strong relationships, you must be attentive, patient, and patient. Remember that each box turtle is unique and not all will tolerate handling. By getting to know your box turtle friend, respecting their boundaries, and creating positive interactions, you can build a strong relationship that promotes their health and well-being. It takes two to tango to build mutual trust, and when you give your box turtle the time and care it needs, you'll find that your bond with it is deeper and more satisfying and long-lasting.

Chapter 5

Enrichment Activities: Keep your box turtle happy and stimulated

Enrichment activities are critical to your box turtle's overall happiness and well-being. Although these reptiles tend to be solitary and lack social activities, they benefit greatly from mental and physical stimulation. In this in-depth examination, we'll explore the importance of enrichment for box turtles, the various enrichment activities, and how to create an engaging habitat that keeps your box turtle content and active.

Recognizing the benefits of improvement

The purpose of enrichment is to create an environment that encourages natural behaviors, mental activity, and physical activity. Box turtles benefit from a well-enriched habitat in the following ways:

1. Psychological Stimulation:

Participatory activities encourage box turtles, prevent boredom and increase their mental well-being.

2. Participating in physical activity:

Participating in enrichment activities encourages movement and discovery, which in turn promotes physical activity—an important aspect of good health.

3. Feature Description:-

When allowed to go about their normal lives, box turtles live a long life. This includes digging, climbing and exploring, among other things.

4. Getting rid of stress:

Stimulating posture reduces stress by allowing natural behaviors to be expressed and preventing the negative effects of boredom.

Many types of enrichment activities

1. Activities related to fodder:

Natural foraging can be done by hiding food in various places around the nest. The turtles become more alert as exploring and searching for food improves their sense of smell.

2. Assess capacity:

Box turtles are burrowers by nature. Provide a sub-section for digging. To mimic their natural habitat, you can dig and bury them in places mixed with dirt, coconut husks or cypress mulch.

3. Climbing Structures:

Although they are not prostrate, box turtles like to climb low buildings. Pebbles, logs, or even small platforms can be added to their enclosures to provide more space to explore and engage in light climbing activities.

4. Properties related to water:

Provide your box turtle with food in shallow water or a small area of water. This allows them to access water and participate in activities such as bathing and crawling that are necessary to protect their skin and shells.

5. Natural elements:

Add natural ingredients to the compost, such as leaves, twigs and non-toxic plants. These elements create hiding places, visual barriers and a more natural, dynamic scene.

6. Novel Products:

Add new items to the enclosure regularly. This category can include balls large enough for the turtle to move around, reptile-safe mirrors, and non-toxic toys. Curiosity and questioning are stirred when something new is revealed.

7. Outdoor Exploration:-

Self-enrichment is greatly encouraged by spending unsupervised time in a safe environment. Exposure to daylight, different textures and outdoor elements can greatly improve their overall well-being.

8. Engaging in Special Relationships:

If you have more than one box turtle, controlled contact can be used for enrichment. But caution must be exercised, and individuals must be warned against any violent behavior.

9. Strong understanding:

Change the environment to stimulate their emotions. New sounds, aromas and textures can be added from time to time. This may include offering different food options to sample or combining safe items with different textures.

10. Circulation of enrichment materials;

Rotate and change enrichment materials frequently to prevent habituation. This preserves the vibrancy of the environment and prevents the turtle from becoming indifferent to them.

Creating a rich environment

1. Perfect Architecture for Containers:

Design the enclosure keeping in mind the natural characteristics of the box turtles. There should be hiding places, a thermally efficient cooking area, and a water source. Use natural materials such as dirt, coconut husks, or cypress mulch to make digging easier.

2. Safe and non-toxic materials:

Whenever new products are introduced, make sure they are made from safe and non-toxic ingredients. Anything with small pieces that can be eaten should be removed. To create a thriving environment, safety must come first.

3. Various foods;

Adapt their diet to include a wider range of foods than they would encounter in the wild. This meets their nutritional needs and adds some fun to mealtime.

4. Feedback and Correction:

Monitor your box turtle's behavior regularly to gauge how well enrichment activities are working for him. Consider bringing things or activities that children are most interested in around them.

5. Outside Cabinets:

If possible, give people a chance to exercise outdoors. Build a safe, well-ventilated outdoor shelter with plenty of natural light. When box turtles are outdoors, they can experience different weather conditions and engage in more vigorous activities.

6. Regularity in daily activities:

Add fun activities while maintaining a consistent daily plan. Because of this, box turtles feel more secure and confident in their surroundings.

7. Close attention to details:

Make sure that treatment brings you satisfaction and happiness. Hosting sessions that involve gentle interaction and allowing them to explore will improve their physical and emotional well-being.

Additional ideas for enriching activities:

1. Teaching Aids:-

Consider including educational toys. To keep their mind engaged while they eat, you can give them puzzle feeders or treat dispensers.

2. Mirror based connection:
Very shy box turtles can be interacted with by placing an object or your palm on the glass. Now they can watch and browse without having to host anything.

3. Circulation of external environment;
If you provide an outdoor enclosure, consider relocating it occasionally. This brings new sights, sounds and smells, creating an ever-changing and exciting outdoor experience.

4. Current differences
Adjust enrichment activities accordingly as seasons change. For example, you can focus on indoor

enrichment during the winter months and increase the amount of outdoor activities during the warmer months.

5. Talking to experts:

Consult veterinarians, reptile behaviorists, or other animal experts for additional recommendations tailored to your specific box turtle's needs. They can provide insight into the species' tendencies and habits.

to sum up:

Enrichment activities are an important part of ethical box turtle care as they greatly increase the overall happiness and well-being of box turtles. By learning about the inner workings of box turtles and incorporating a variety of stimulating activities into their daily routine, it is possible to create an environment that expresses their physical well-being, mental stimulation, and natural behavior. Remember that each box turtle is individual and has different preferences. Notice their reactions, adjust your actions accordingly, and experience the complete joy of providing your box turtle friend with a happy and fulfilling life.

Chapter 6

Social creatures: interpretation of behavior and communication in box turtles

Understanding the behavior and communication methods of box turtles is critical to providing the best care and building a stable relationship with these amazing reptiles. Although not naturally social animals, box turtles exhibit unique behaviors and ways of communicating that reflect their needs, wants, and responses to their environment. In this comprehensive analysis, we examine box turtle behavior, communication cues, and the importance of understanding their behavior to ensure a happy and healthy life.

Understanding how box turtles work:

1. Execution within limits:

Box turtles are well known for their territory. They establish home ranges and defend their environment in the wild. In captivity, they can become aggressive towards other box turtles. Adequate room and resources should be available to minimize territorial disputes.

2. Sleepiness

During the winter, box turtles naturally hibernate. They can show behavioral changes, such as an increase in burrowing tendencies and a decrease in activity, when the temperature drops and the number of daylight hours decreases. A suitable sleeping place should be provided for their safety.

3. Body temperature control and baking;

Box turtles need to be in the sun to survive. In order to maintain a healthy body temperature, they need to find an oven with a suitable thermal efficiency. This function aids in metabolism, vitamin D production and digestion.

4. Burying and Diving:

Because of their natural propensity for burrowing, box turtles in captivity may burrow for the purpose of exploring, finding hiding places, or nesting. When a substrate is available for digging, such as dirt or coconut husks, their natural tendencies are encouraged.

5. Viewing and Collection:

Box turtles are inquisitive creatures that enjoy exploring their environment. When foraging, they use their keen sense of smell to help them find food. Their mental health increases when they are encouraged to ask questions by various environmental signs.

6. Seeking security:

Box turtles seek shelter when the weather is bad or when they sense danger. Encouraging them to have half-log structures, shelters, or hiding places will make their confinement more secure.

7. Bathing and tanning;

Basking and basking are important traits for box turtles. Exposure to natural sunlight promotes the production of vitamin D and improves their physical health. By providing a small bowl of water, you encourage dipping habits that improve the shell's hydration and health.

Interaction symbols:

Box turtles do not communicate in the same way as social animals, but they communicate through a variety of signs and behaviors. Identifying these communication signals is critical to ensuring their interest and establishing a happy, stress-free environment.

1. Revealing and retreating:

When a box turtle retreats or hides behind its shell, this may indicate that it needs some alone time or is reacting to danger. At this time, respecting their request for isolation is important for their safety.

2. In slow motion:

Box turtles often move purposefully and slowly. Irritation or rapid movements can indicate discomfort, anxiety, or a defense mechanism against something perceived as threatening. You can assess their overall health by looking at how they move.

3. Movement of hands, feet and head;

Box turtles communicate through their delicate movements. A person may nod or nod their head in recognition or in response to external stimuli. Leg movements that involve extending or returning the legs may indicate discomfort or discomfort.

4. It says:

Box turtles are not vocal like other reptiles, but they do hiss when they are scared or uncomfortable. This can be a defensive move, so it's important to talk to them gently to ease the tension.

5. Feeding and Appetite Symptoms:-

Their eating habits provide insight into their overall happiness and well-being. A healthy box turtle that is adequately fed will be interested in eating, but an unexpected decrease in appetite may indicate underlying issues that need to be addressed.

6. Favorites for bathing:

Box turtles communicate their preferred temperatures through their basking behavior. Frequent basking shows that they are regulating their body temperature, but avoiding the heated area can be a sign of discomfort or displeasure with the temperature difference.

7. Wet conduct

One way to show that you need water is by drinking. If you give them a shallow bowl of water, you can adjust how hydrated they are, and you can see how they're drinking, so you can be sure they're getting enough water.

8. Areas of Anger and Protection:-

Territorial disputes or perceived threats may lead to defensive postures. Box turtles may display head movements, limb extensions, or hissing sounds as signs of aggression. Reducing aggressive behavior can be achieved by identifying people or taking action against environmental stressors.

Environmental factors that affect behavior:

1. Temperature and lighting;

The temperature and light conditions in the habitat have a great influence on the movement of the box turtle. Inadequate temperature or UVB exposure can lead to stress and health problems. Adequate heating and lighting systems are essential for their safety.

2. Enclosure Size and Position:-
The size and location of the enclosure affects the box turtle's behavior. Inadequate space can lead to stress and territorial disputes, but well-designed areas, hiding places, hot spots and suitable surfaces encourage natural behavior.

3. Social organization;
Social activity in multiple turtle habitats influences behavior. Dominance tendencies, gender roles, and territorial disputes can all cause anxiety. Giving your box turtles plenty of space and knowing their social

dynamics are important to creating a stable living environment.

4. Existence of difference:-

The presence of other box turtles can affect the behavior. Generally, box turtles are solitary animals, but occasionally form social bonds with others or exhibit territorial behaviors. By monitoring their relationship and providing adequate supplies, it will reduce the stress that may occur.

5. Current differences

Box turtles may exhibit seasonal changes in their activity, such as reduced activity in the winter and increased foraging in the summer. They can maintain their biological cycle by mimicking the seasonal variations in their natural environment.

Developing Relationships by Observing:

1. Frequent Correspondence:

Building trust requires polite yet formal communication. Spend some time in the enclosure, talk to them in a soothing voice, and observe their behavior. Gradual exposure to handling promotes a strong bond if they feel safe.

2. Hand feeding:

Feeding someone by hand can encourage positive associations. Offer them the food they prefer so they associate your presence with positive things. This builds trust and reduces anxiety associated with handling.

3. Regular Observation:-

Monitor your box turtle's movements regularly. Check regularly for changes in stress, hunger, and activity level. Timely intervention is possible through early detection of behavioral change.

4. Respect for others' personal space

It is important to respect their request for privacy. Their comfort is assured when they can retreat to hiding places or enjoy isolation while engaging in certain behaviors, even though they still need contact.

5. Important incentive

A snack or other positive reinforcement can be used to reward the desired behavior. They are more likely to repeat positive behaviors when they receive praise or show interest in an enrichment item.

Common behavioral problems:

1. Actions related to anxiety:

Many symptoms can be indicators of stress, including appetite, fatigue, or defensive body language. Recognizing and managing stressors such as inappropriate enclosure conditions or environmental changes is critical to their safety.

2. Anger between several turtles:

Where box turtles are abundant, territorial disputes or aggressive behavior may occur. Hostility can be controlled by working in multiple locations, monitoring social changes, and separating people if necessary.

3. Whether or not to request:

Decreased curiosity or inactivity may be symptoms of environmental dissatisfaction or health problems. Ensuring comfortable living conditions, providing a variety of enrichment and protecting against health risks are critical.

4. Reluctance to eat;

When a person suddenly stops eating, they need help. It can indicate eating disorders, physical issues or stress.

To rule out underlying medical issues and make dietary or environmental improvements, it is important to see a veterinarian.

to sum up:

To truly understand the behavior and communication of these dynamic creatures requires patience, observational skills, and a deep understanding of box turtle characteristics. By interpreting their emotions, respecting their natural tendencies, and creating an environment that promotes their well-being, you can have a happy and rewarding relationship with these amazing reptiles. A thriving and contented companion box turtle is the result of attention to detail, consistent monitoring and habitat adjustments. As you begin your journey to understand their behavior, you'll find that every interaction you have with them provides valuable insight into the fascinating world of box turtles.

Chapter 7

Species Highlights: Analyze the different types of box turtles for pets

Box turtles are amazing reptiles that belong to the terrapin genus and are known for their unique characteristics and appearance. Because of their distinct habits, colors, and shell patterns, box turtles are a popular choice for pets. In this comprehensive review, we'll look at several box turtle species that can be kept as pets, highlighting unique characteristics, care requirements, and things owners should know about each species.

Turtle: Eastern box turtle, Terrapene carolina carolina

Physical Characteristics:

The eastern box turtle is distinguished by the complex patterns on its domed carapace, which range from orange to brown in color. Because the turtle has a long, mobile shell, it can completely surround itself for protection. They often have beautiful colors on their heads and legs, such as orange, yellow or red.

region and natural environment;

The eastern United States is home to eastern box turtles, which live in a variety of habitats such as forests, meadows, and grasslands. They grow in areas with lots of vegetation and shallow water and easily accessible land.

Business Care;

- When setting up the enclosure, make sure it is wide and full of leaf litter, litter, and litter mix. Include a shallow dish for diving in water and other hiding places.

- The diet of eastern box turtles is varied. Make sure your pet eats a variety of foods, including fruits, vegetables, worms, and insects.
- Temperature and lighting: Use UVB lighting and maintain temperature levels for optimal shell and bone health.
- Interaction: Although not as friendly as other reptiles, eastern box turtles can show signs of curiosity and curiosity towards their owners. Respect their request for privacy and avoid too much interference to ease tensions.

Three-toed Box Turtle, Terrapene carolina triungis:

Physical Characteristics:

Similar to the three-toed box turtle, the eastern box turtle can be identified by having three toes on each hind foot. The carapace has many colors such as olive, black and brown. The scale of the toes is an important indicator.

region and natural environment;

Parts of the central United States and Mexico are included in the range of the three-toed box turtle. In the open forest and in the fields, loose, sandy soil, it is more widespread.

Business Care;

- Create a hole covered with sand that is heavily planted. Include hiding places and a shallow water bowl for drinking and bathing.
- Diet: Three-toed box turtles will eat everything. Offer a variety of tasty foods, including fruits, vegetables, insects, and snails.
- Temperature and Illumination: Provide thermal efficiency and UVB lighting. These elements are

important for thermoregulation and calcium metabolism.

- Interaction: While they may display identification marks, three-toed box turtles are similar to eastern box turtles in that they require little to no interaction. To avoid stress, make sure everything is peaceful.

The sophisticated box turtle, Terrapene ornata:

Physical Characteristics:

The most distinctive feature of the ornate box turtle is its long domed shell, which is dotted with radial golden lines. The shell has different colors like orange, black and brown. In general, males show brighter colors than females.

region and natural environment;

Native to the central United States, ornate box turtles live in open woodlands, grasslands, and meadows. They grow in sandy soil conditions.

Business Care;

- Setting up the enclosure: To mimic their natural habitat, use sandy soil or a sand-soil mixture. Provide hiding places and a shallow water bowl for diving and drinking.
- Food: Ornate box turtles are omnivorous but prefer a variety of plants, insects and snails.
- Temperature and Lighting: Maintain the right temperature and provide UVB lighting for optimal health. It is important to heat the place for temperature control.
- Contact: Limit your contact with ornet box turtles to avoid stress, although they are more tolerant of handling than other species. Give them plenty of hiding places so they feel safe.

Gulf Coast Turtles, or Terrapin Carolina Major:

Physical Characteristics:

The Gulf Coast box turtle is a subspecies of the eastern box turtle, characterized by its highly domed carapace and variable coloration. They may have vibrant patterns and motifs on their heads, feet, and shells.

region and natural environment;

As the name suggests, Gulf Coast box turtles are found along the Gulf Coast of the United States. They prefer open and forested areas with water supply.

Business Care;

- Create an enclosure by creating habitat, foliage and hiding places from the soil. Include a small bowl of water for dipping and rinsing.

- Food: Gulf Coast box turtles are omnivores and eat a wide variety of fruits, vegetables, worms and insects.
- Temperature and Illumination: Provide thermal efficiency and UVB lighting. Areas exposed to the sun are very important for temperature control.
- Interactions: Although Gulf Coast box turtles tolerate occasional handling, their stress levels should be monitored. Allow them plenty of time to explore on their own.

Western Box Tortoise, Terrapene ornata lutola:

Physical Characteristics:

The Western box turtle, a subspecies of the Ornet box turtle, is distinguished by the unique radial lines on its shell. They are orange, yellow, black and brown. Males are more colorful than females.

region and natural environment;

In the western United States, prairies, forests, and desert areas are home to western box turtles. They grow in sandy or well-drained soil.

Business Care;

- Setting up the enclosure: To mimic their natural habitat, use sandy soil or a sand-soil mixture. Provide hiding places and a shallow water bowl for diving and drinking.
- Food: Although they are omnivores, Western box turtles enjoy a variety of plants, insects and snails.
- Temperature and Lighting: Maintain the right temperature and provide UVB lighting for optimal health. Areas exposed to the sun are very important for temperature control.
- Engagement: Like other box turtles, western box turtles can tolerate handling, but prefer little to

no engagement. Create a peaceful space with hiding places to reduce stress.

Consideration for potential owners

1. Legal issues to be addressed:-

Before making a purchase, review and understand all applicable national and international laws regarding box turtle ownership. Some species may require a permit as they may be protected.

2. Life span:

Box turtles are long-lived - some have been known to live for decades. Prospective owners should be prepared for a long-term commitment and the responsibilities that come with it.

3. Space Requirements:

Box turtles need a lot of space to explore. Make sure the enclosure is spacious, lush with greenery and full of hiding places that suit their natural tendencies.

4. Essential Ingredients:

It is important to understand the nutritional needs of certain box turtle species. To be healthy, look for foods rich in fruits, vegetables, nuts, and calcium supplements.

5. Veterinary treatment:

Regular animal check-ups can provide preventive care and early detection of potential health conditions. Find a veterinarian who specializes in treating reptiles.

6. Environmental improvement;

Include areas for sunlight, hiding places, and exploration to enhance the enjoyment of the enclosure. Rotate them regularly and add new boosters to keep their brains active.

7. Temperature control;

Keep the temperature levels in the enclosure suitable. Box turtles' digestion, metabolism, and general health depend on external heat sources.

8. Observing how things normally work

The natural characteristics of box turtles, such as solitude and little treatment, should be respected. Provide an environment where people are free to work on their passions.

to sum up:

For reptile lovers, owning a box turtle can be a fun and rewarding experience. Each breed brings unique characteristics, colors and ways of behaving to the table, giving future owners many options. Understanding the specific needs of box turtle species is important to ensure their longevity and safety. Whether it's the striking colors of the ornate box turtle, the unusual number of toes on the three-toed box turtle, or the delicate patterns on the eastern box turtle, each species brings something unique to the different box turtles. A happy and fulfilling relationship with these amazing reptiles is facilitated by providing them with good care, an engaging environment and recognizing their individuality as conscientious pet owners.

Chapter 8

Grooming Tips: Keeping your box turtle shell clean and healthy

The shell of the box turtle is an integral part of the body, it serves as a defense and a measure of general health. Maintaining the health of these amazing reptiles depends on taking proper care of their shells. This comprehensive guide covers grooming techniques designed to provide your box turtle companion with a strong, clean, and healthy shell.

Understanding the Box of Tortoise Shells:

1. The importance of shell

Box turtles are unique because they have a shell that protects them from the environment and predators. The two main parts are the plastron or lower shell and the carapace or upper shell. The condition of the turtle's keratin-covered, bony shell serves as a reliable barometer of its overall health.

2. Expansion and Development of Shell:-
A box turtle's shell changes dramatically as it grows. A newborn turtle's shell is soft and fragile, but as it ages, it gradually hardens. A balanced diet, UVB exposure, and environmental conditions all contribute to the formation of a healthy shell.

3. Shell Issues and Differences:
Box turtles may have cracks in their shells, deformities, pyramiding, or decaying shells. Pyramiding, also known as elevated shell scutes, is a typical consequence of malnutrition and dehydration. Environmental factors or unsanitary conditions can cause bacterial or fungal diseases, including shell rot.

Tips to keep your shell healthy:

1. Shell Health Diet:

Maintaining a healthy shell requires a balanced diet. Make sure your box turtle eats a variety of foods that include leafy greens, insects, worms, and calcium-rich foods. Calcium is important for the growth and strength of the shell.

2. Drink a lot of water

Water consumption is critical for healthy shells. Provide a bowl of shallow water in the enclosure for people to drink and wash. Because box turtles absorb water through the epidermis, they generally stay hydrated and maintain the integrity of their shells.

3. UVB rays and sunlight;

Exposure to natural sunlight or UVB light is essential for vitamin D production. This is important because calcium needs vitamin D to absorb. Make sure your turtle gets some sunlight or UVB light exposure every day.

4. Keeping the fence clean:-

It is very important to keep things clean to prevent shell problems. It is important to clean the enclosure regularly to remove any material that may hold moisture, debris or uneaten food. Unclean or damp materials can become a home for fungus or bacteria.

5. Regular Medical Tests:

Visually inspect your box turtle's shell regularly. Check the area for any changes in the emergency, colors, breakage or strangeness. Finding a problem early allows for quick intervention.

6. Take good care of:

Show respect and kindness to your box turtle. Avoid unnecessary force or compression as this may damage the shell. When lifting your turtle, be sure to support the plastron and carapace to prevent injury.

7. Avoid pyramiding

In box turtles raised in captivity, pyramiding is a common issue, usually associated with malnutrition and dehydration. Provide nutritious food, avoid overfeeding and ensure everyone has access to clean water to prevent pyramiding.

8. Dealing with Shell Abnormalities: As soon as you notice any abnormalities in your box turtle's shell, take him to a reptile vet immediately. Defects may indicate inappropriate environmental conditions, metabolic issues, or inadequate nutrition.

9. How to handle Shell Rot

Shell decay needs immediate attention. If you see any symptoms, such as soft spots, discoloration, or an unpleasant odor, consult a veterinarian. Treatment may include topical antifungal or antibacterial medications, improved living conditions, and improved hygiene.

10. Avoidance of faults and damages;

Provide a safe shelter for the shell to prevent stress and injury. Sharp objects, uneven surfaces and areas with insufficient concealment can all be considered hazards. Reduce the chance of accidents to maintain the integrity of the shell.

Detailed decorating method:

1. Observation and Evaluation:

To begin, carefully examine the shell of the box turtle. Look for any signs of an emergency, such as discoloration or unusual changes. Note areas that need further investigation.

2. Easy cleaning;

The surface of the shell should be cleaned with a damp cloth or sponge. Avoid sharp objects and harsh chemicals as they can damage the keratin layer of the skin. Make sure the fabric is free of any particles that could damage the shell.

3. Drowning

Provide a shallow bowl of water for your box turtle to relax in. In addition to moisturizing, watering can also soften any materials or debris stuck to the shell. Let the turtle drink for fifteen to twenty minutes until it feels safe.

4. Cleaning if necessary;

To remove stubborn debris or algae, use a soft-bristled toothbrush. Gently wipe the shell, paying special attention to the regions where the debris has accumulated. Try to move in a gentle circular motion with minimal force.

5. Cleaning and Cleaning:-

After washing and cleaning, rinse the turtle shell with clean water to remove any residue. Dry the peel with a soft and clean towel. Make sure the turtle is thoroughly dry before returning it to the enclosure.

6. Drying the Shell Skies:

To maintain healthy squibs, you can add a thin layer of conditioner or an oil specifically formulated for reptiles after drying. This can moisten the squids and preserve the natural light of the shell. Avoid over-applying because over-applying can attract dirt.

7. UVB Pollution:

Place your box turtle in an area that receives direct sunlight or make sure it has access to UVB lights as recommended. This exposure increases the synthesis of vitamin D, which is important for shell health and calcium absorption.

8. Regular trips to the vet

To ensure your box turtle's overall health and well-being, schedule regular veterinary checkups. A veterinarian experienced in treating reptiles can perform a thorough examination, provide dietary recommendations, and manage any emerging health problems.

Preventive measures to maintain shell health:

1. Balanced diet

Provide a balanced diet that meets all of your box turtle's nutritional needs. Consult a veterinarian who specializes in pet animals to develop a specific feeding plan for your turtle's age, breed, and overall health.

2. Taking calcium supplements:

Administer calcium supplements as prescribed by your veterinarian. Calcium is important for the strength and integrity of the shell. Make sure the supplements are suitable for reptiles and follow the suggested dosage.

3. Sufficient moisture;

To ensure proper humidity, make sure the enclosure has a shallow bowl of water. Check the cleanliness and hygiene of the water to encourage consistent drinking and drinking habits.

4. Perfect conditions for incubation:

Provide perfect enclosure conditions including thermal efficiency, hiding places and suitable flooring. Minimize stressors such as sudden temperature changes and lack of hiding places.

5. Environmental improvement;

To encourage natural features, improve the surrounding environment. Include things like hiding places, sunbathing areas, and other things you can think of. Participating in stimulating activities increases mental stimulation and overall well-being.

6. Continuous Cleaning Time Table:

Provide a standard procedure for cleaning the enclosure. Remove leftovers, don't eat anything and keep a clean slate. Clean any wet or dirty areas once in a while to prevent shell problems.

7. Avoid Overcrowding:

Although box turtles require some handling, it is important to keep their stress level as low as possible. Avoid giving the turtle too much attention, especially if it appears sick or uncomfortable.

8. Rapid veterinary treatment;

If you notice any changes in your box turtle's shell, behavior, or general health, contact your veterinarian at once. Early detection and intervention are critical to addressing potential problems.

Common health problems associated with Shellac and their medications include:

1. Pyramiding

- Problem: Pyramiding is often associated with malnutrition and dehydration, and promotes squatting.
- Intervention: Adjust diet to ensure proper nutrient intake. Increase the amount of fresh water available to maintain hydration. Consult your veterinarian for more detailed instructions.

2. Changing scales;

- Problem: Soft spots, discoloration, or an offensive odor can be caused by a bacterial or fungal infection, such as shell rot.
- Intervention: Consult a veterinarian for diagnosis and treatment. Topical antifungal or antibacterial medications may be prescribed. Improve hygiene and living conditions.

3. Injury or damage:

- Problem: Accidents or rough handling can damage or damage the shell.
- Intervention: Create a safe environment to reduce the chance of error. Consult a veterinarian for guidance on the best course of action and shell repair in the event of an injury.

4. Abnormalities:-

- Problem: Shell abnormalities can result from inadequate nutrition, metabolic conditions, or environmental factors.
- Intervention: Consult a veterinarian for guidance on identifying and treating underlying problems. Adjust the diet, improve the environment and follow the advice of the veterinarian.

5. Excessive drinking;
- Problem: Growing spurs can cause discomfort and inhibit proper shell development.
- Intervention: Using appropriate equipment, carefully trim large scutes under veterinary supervision. Maintain a balanced diet and adequate hydration to promote proper shedding.

to sum up:

Caring for your box turtle and maintaining a clean, healthy shell are the first steps in responsible reptile care. By understanding your turtle's unique needs, providing a balanced diet, keeping him hydrated, and providing frequent care, you can improve his overall health. Maintaining the health and strength of your box turtle friend's shell requires regular veterinary examinations, prompt action when issues arise, and a commitment to providing a rich habitat. The time and effort you put into keeping your box turtle shell healthy will enhance your bond with them as you embark on this caring adventure.

Chapter 9

The life cycle of the box turtle: from hatchling to adulthood

The life cycle of a box turtle is a fascinating journey consisting of several developmental stages, each characterized by unique challenges, unique characteristics, and critical moments. Understanding the life cycle of these fascinating reptiles is critical to providing them with optimal care and ensuring their health. In this in-depth investigation, we follow the box turtle's remarkable journey from hatchling to adulthood.

1. Egg stage;

- Egg laying and nest building;

The nesting and egg-laying period is the beginning of the box turtle's life cycle. Female box turtles require well-drained soil when building their nests. This kind of thing usually happens in spring or early summer. The female carefully deposits a clutch of eggs into a hole she digs with her hind legs. The number of eggs varies depending on the species, although it usually ranges from one to nine.

- Time until symptoms appear:

After hatching, the eggs must go through an incubation stage that lasts for several weeks. Two factors that affect the length of care are temperature and humidity. Warmer temperatures generally lead to a shorter incubation period. The eggs are given space to grow and hatch organically.

2. Stage of development;

- Coming from Nest:

As the incubation period comes to a close, the chicks begin to leave the nest. Two frequent environmental cues that trigger this phenomenon are heat and brightness. Chicks use the egg tooth, a small projection on the front of their upper jaw, to open the egg shell.

- Survival and Vulnerability Problems:

At this stage, chicks are very weak. Some of the challenges they face include finding predators, dangerous environments, and suitable hiding places. Many box turtles lack parental care, so hatchlings must rely on instinct and their own abilities to navigate their environment.

3. First step:

- Shell and expansion development;

When box turtles hatch, they go through a juvenile period. At this stage their shells are flexible and delicate, and become harder as they grow. Adequate nutrients, UVB sunlight and suitable habitat are essential for the formation of healthy shells. Their shells have distinctive patterns that become apparent over time.

- Change diet and feed;

Adolescent box turtles actively seek out a variety of food sources, including small insects, worms, snails, and plant material. As they grow older, their diet changes, transitioning from being primarily insectivores to omnivores and eating plant- and animal-based foods.

- Salvage potential:

Young box turtles are still vulnerable to predators at this stage. Their small size and thin shells make them vulnerable to predators such as mammals, birds and other reptiles. Finding suitable hiding places will be critical to their survival.

4. Sub-Level:

• Maturity and gender differences;

As box turtles progress through the subadult stage, they experience sexual differentiation and maturation. Males and females can be distinguished from each other because sexual characteristics are more visible. Male placentas are usually larger and flatter, while female placentas are flatter.

• Ethical and territorial disintegration;

Young box turtles may exhibit territorial behavior by establishing and maintaining their natal territories. Some people may start dispersing in search of a mate or a suitable environment. This stage is a significant turning point on the road to adulthood.

- Maintaining growth and shell formation;

In the sub-adult stage, box turtles continue to grow and their shells eventually harden. It is still necessary to find adequate food and suitable environments for general health and to prevent abnormalities in the shell.

5. Level of maturity:

- fully grown seeds;

The adult stage is characterized by reproductive maturity. Once they reach adulthood, box turtles have the ability to breed and engage in an annual nesting cycle. Matrimonial traits include courtship displays in which males actively seek out females and engage in special rituals to attract mates.

- Nesting in the reproductive cycle;

Adult female box turtles follow reproductive cycles, searching each year for suitable nesting sites in which to lay a clutch of eggs. There are two environmental factors that usually influence this behavior: temperature and daylight duration. Males actively seek females during breeding season.

- establish limits;

Adult box turtles expand territories that they can defend against invaders. Territorial behaviors include head nodding, vocalization, and angry posture. It is normal for a healthy population to need enough room and housing to support many individuals.

- Longevity and longevity;

Box turtles are known for their longevity as some can live for decades. The life span of a box turtle is determined by many factors such as heredity, environment and level of human care. Regular veterinary care and good husbandry will result in a long and healthy life.

Factors affecting the life cycle

1. Environmental Conditions:-

Throughout their various life cycles, temperature, humidity and daylight duration are critical factors for box turtles. These environmental conditions affect the nest-building process, the length of the incubation period, and the time at which the chicks hatch.

2. Quality of Residence:

The quality of the habitat has a great influence on the life cycle of the box turtle. All life stages of box turtles have access to a variety of fresh water sources, appropriate nutrition, and hiding places.

3. Effects on people:-

Human activities such as habitat loss, pollution, and contact with cars and roads pose serious threats to box turtles. Conservation efforts, habitat protection, and ethical interactions with box turtles are essential to reduce the impact of human activity.

4. Predation and Survival Challenges:-

Throughout their lives, box turtles are exposed to a variety of predators. Predators, including raccoons, foxes, birds of prey, and snakes, can affect the survival of juveniles, juveniles, and adults. Making adequate hiding places and reducing the chances of predators are two important things for their survival.

5. Reproductive success:

Breeding success is influenced by the availability of mates, suitable nest sites and the general health of the individual. Human movement, habitat fragmentation, and climate change can all affect population movement and reproductive success.

Thinking about protection

1. Protection of the natural environment;

Protecting the native habitats of box turtles is important to their conservation. Protecting natural areas, reducing habitat degradation, and creating wildlife corridors all contribute to the overall health of box turtle populations.

2. Highway Intersections and Fatalities:

Road mortality poses a serious threat to box turtle populations, especially during breeding and dispersal periods. The harmful effects of roads on box turtle populations can be minimized by implementing programs such as fencing, wildlife crossings, and public awareness programs.

3. Initiatives for the rehabilitation and release of wild animals:

For the treatment of injured or lost box turtles, wildlife rehabilitation centers are essential. Rehabilitation programs aim to allow people to conserve wild populations and return them to their natural habitats when appropriate.

4. Observation and Population Science:

By participating in citizen box turtle control projects, valuable information can be learned about population health, distribution, and potential threats. The ecology of box turtles is better understood thanks to citizen science initiatives that provide direction for conservation efforts.

5. Education and Dissemination:

Box turtle conservation relies on public education about the species' importance, life cycle and challenges. Outreach programs have the potential to foster awareness and encourage action to protect box turtles.

to sum up:

A fascinating journey through the life cycle of a box turtle through evolution of improved reproductive strategies, challenges to survival and adaptation. From the helpless goat emerging from the egg to the mature adult capable of transmitting information to the next generation, each stage demonstrates the resilience of these amazing reptiles. Because we are the stewards of the ecosystem, the conservation of the box turtle depends on people. If box turtles are to thrive in the wild for many generations, conservation initiatives, habitat protection, and respectful interactions are essential. As we learn more about the life cycle of these exotic animals, we grow to respect them and the environment in which they live.

Chapter 10

Box Turtle FAQ: Frequently asked questions and myths addressed

Because of their unique appearance and fascinating habits, box turtles are a popular choice for companion reptiles. However, they require special attention and care just like any other pet. In this comprehensive guide, we address common concerns and misconceptions about box turtles in an effort to provide accurate information and promote responsible ownership.

1. Can turtles be petted?

Yes, but with a few disclaimers:
Box turtles can make wonderful pets if their owners are prepared to invest the time and effort required in proper care. However, it is important to understand that

they require specific environments, a balanced diet and regular vet care. Because of their longevity and changing care needs as they age, box turtles require a long-term commitment from their owners.

2. Which box turtles make good pets?

Favorite pets:
Of all the box turtle species kept as pets, the eastern box turtle (Terrapene carolina carolina) is the most popular. The ornate box turtle (Terrapen ornata), the Gulf Coast box turtle (Terappen carolina major), the three-toed box turtle (Terappen carolina triungis), and the western box turtle (Terappen ornata lutola) are among other species. He was held captive. It is important to research the species and choose the ones whose requirements match your ability to take care of it.

3. Do box turtles need a lot of space?

Yes, they need enough space:

To thrive, box turtles need a large, well-protected environment. If placed indoors, the enclosure should have adequate space for inspection, as well as hiding places, an escape hatch, and appropriate padding. The best enclosure is outdoors. The size of the enclosure depends on the type and number of turtles. As long as it meets their specific needs, the bigger, the better.

4. How should I feed my box turtle?

A variety of foods are important:

Box turtles are omnivores and require a variety of foods. Offer a variety of worms, snails, fruits, vegetables, and insects (such as mealworms and crickets). Calcium supplements may be especially important for caged turtles. Avoid feeding your dog or cat a high-protein diet on a regular basis, as this can lead to nutritional imbalances.

5. Can multiple box turtles be kept in one enclosure?

It should be carefully considered:
While some tortoises get along well with other tortoises, it is important to take extra care when keeping multiple tortoises in an enclosure. Aggressive and territorial behavior may occur, resulting in conflict or injury. Finding hiding places, creating enough space, and monitoring interactions are important. Be prepared to split the turtles if any disagreements arise.

6. Do box turtles need UVB lighting?

UVB is really important.
Box turtles require UVB sunlight to produce vitamin D, which is essential for healthy calcium absorption and overall well-being. Allow daylight into the enclosure or use a UVB light to illuminate it. As the UVB output

decreases over time, be sure to replace the lamp according to the manufacturer's recommendations.

7. Can I handle my box turtle more often?

The best treatment is minimal:
Although some box turtles tolerate handling, they usually prefer minimal interaction. They can become defensive or self-absorbed when touched too much. When handling is necessary, be careful not to injure the animal by supporting both the plate and the carapace.

8. Do box turtles need water to swim?

Sufficient water
Box turtles require access to shallow water for drinking and bathing, but do not need deep water for swimming. Serve with a shallow dish so they can be easily absorbed.

Make sure they can get in and out of the bowl and that the water is always clean.

9. Can box turtles live inside?

Specifically, given the correct configuration:
Box turtles can live indoors with the right conditions. Make sure the enclosure has adequate hiding places, a UVB-lit enclosure, the right amount of nutrients, and a balanced diet. The size of the enclosure should satisfy the turtle's curiosity. Indoor environments require frequent care and cleaning.

10. Do box turtles ever hibernate?

Some species are dormant.
Shredding is a stage that many box turtles go through, comparable to hibernation. During this period, they are inactive and their metabolic rate decreases. Different

species have different requirements for sleep, and not all box turtles hibernate. For those who do, it is important to mimic seasonal variations in nature.

11. How can I tell the sex of a box turtle?

The physical characteristics vary:
One way to determine the sex of a box turtle is to examine its physical characteristics. Males usually have longer, thicker tails, larger, more convex plates, and sometimes more colorful patterns. In general, females have a short, thin tail and a flat plate.

12. Is salmonella carried by box turtles?

It is advised to be careful-
Box turtles, like other reptiles, can carry the Salmonella bacteria. It is very important to maintain proper hygiene when working with them. Avoid contact with their

habitat or water, and always wash your hands after handling. Be especially careful around young children, the elderly, or people with weakened immune systems.

13. What does box turtle pyramiding involve?

Improper care can cause shell damage;
The pyramiding of the shell is known for its high success rates. In the first years of a turtle's life, malnutrition and dehydration are often the cause. Pyramidine can be avoided by eating a balanced diet, ensuring access to clean water, and maintaining proper hydration.

14. I have a captive box turtle; Should I release it into the wild?

It is not recommended to release captive turtles:
In general, it is not a good idea to release caged box turtles into the wild. In addition to losing essential life

skills, captive-bred turtles can transmit diseases that can affect wild populations. Additionally, releasing local laws without adequate consideration may result in legal repercussions.

15. How old are box turtles?

Extended life span:

One characteristic of box turtles is their long lifespan. They can live for decades in captivity; Some can live for fifty years or more. Their overall health and lifespan is controlled by proper care, food and regular veterinary care.

16. Do box turtles hibernate?

Every day with some exceptions:

Because of their diurnal nature, box turtles are typically active during the day. Although there may be differences between people and species. Between dawn and dusk, some may be more active, while others may show more nocturnal behaviors. In captivity, maintaining a constant cycle of light and dark can help regulate their movements.

17. Do box turtles shed shells?

Shells are not true.

Despite popular belief, box turtle shells are not discarded. Like some reptiles, whose skin is shed, their shells do not shed completely, although the scutes on them gradually wear away and may be replaced over time.

18. Can I replace my box turtle with sand?

Be careful when working with sand:

Sand can be part of the subsoil, but use caution when doing so. Small particles of sand have the potential to cause impact, which can be dangerous. A more suitable environment can be created by combining sand with other materials such as cypress mulch or coconut husks.

19. Do box turtles have anger issues?

Regional actions can be taken:
Box turtles are known to exhibit territorial behavior, especially during spring planting. Although they are not aggressive by nature, conflicts can arise if many turtles are kept in a small space. It is important to provide suitable hiding places and monitor their movements.

20. Can I paint my box turtle's shell?

Totally not recommended.

Painting the shell of a box turtle is strongly discouraged. An important part of their structure is their shell, and any substances applied to it can be harmful. Color can block important UVB rays and lead to health problems. Never prioritize their physical health over beauty choices.

to sum up:

Understanding their needs and behaviors is essential to providing proper care for box turtles and building a close relationship with these amazing reptiles. By dispelling common myths and answering frequently asked questions, we empower both existing and new owners to make decisions that support the health and well-being of their box turtle companions. By practicing responsible ownership and committing oneself to ongoing education and adaptation, a positive and meaningful experience can be ensured for both humans and box turtles.